Discover And Contact Two

IAN BEARDSLEY

CHAPTER 1

Other Half

The author of Discover And Contact has realized that having told only half the stories of his experiences in the Sacromonte that there must exist again as much as was discovered there.

People in the Sacromonte came to think I was interested in a Gypsy woman named Belen (Bethlehem). They may have been right.

The son of Manuel, the Gypsy Shaman, was someone who was teaching me the principles of flamenco. His name was Antonio and he was living with a Spanish woman, Tamara, in an apartment provided to him by his uncle, a restauranteur to use while he pursued a career in Flamenco. Tamara told me that if I married a Gypsy woman and she saw me talking to another woman, she would scratch her eyes out.

One day Antonio and I were listening to Paco De Lucia, his album Sirocco at his apartment. At some point he told me Camaron was a great singer because he had a good waiver and he signaled an oscillating motion up and down in his throat with a waving of his hand.

At some point he also told me he was learning the syncopated bulerias (bulerias being a style of playing in flamenco).

I think now, of course, syncopation. How do you superimpose three over four such that each of the three beats are equally spaced and each of the four beats are equally spaced. You play both on the one, a beat on the upbeat of two (the two and), and both on the four. Graphically that is:

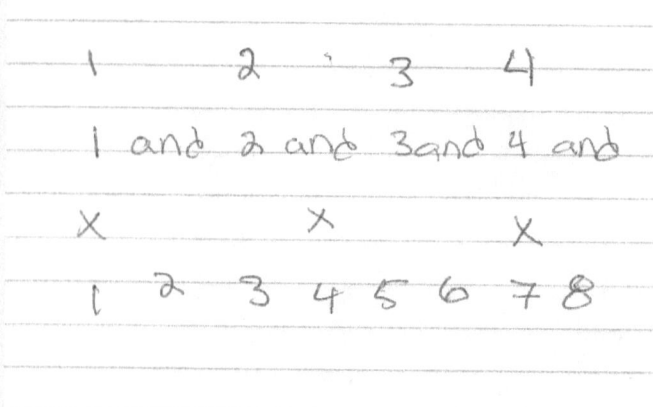

That gives the three numbers $(1, 4, 7)$. In my first cosmology, in my work Contact And Discover, the three important numbers were $(9/5, 5/3, 2)$. $9/5$ is the yin of the universe, $5/3$ is the yang of the universe, and 2 is the taoistic idea of the Gypsy Shaman Manuel that is "God is an idea, there are two elements". We have then, two sets, $(9/5, 5/3, 2)$ and $(1, 4, 7)$. The new set is quite distinct from the first. What planets do they represent? Answer is:

Mercury (1)
Mars (4)
Uranus (7)

$7+4+1 = 12$

Their sum is twelve and 12 is the meter of flamenco.

Ian Beardsley
April 10, 2014

Opening Doors

When we think of blues guitar we play 4 beats in the base and three in the melody or vice versa. We can think of this as 3 super-posed over 4. We consider this for the three section as 1, 2, 3, ... 1, 2, 3,... 1, 2,... The last group of beats on the three only gets two beats. So this adds up to six and two thirds. The four section is four beats. So we compare 6 and 2/3 to 4 and get:

20/3 divided by 4 = 20/12 = 10/6 = 5/3

Thus it is an expression of five thirds (5/3).

We have seen this before, 9/5 was the yin of the universe and, 5/3 was the yang. This comes from the biological or five-fold symmetry:

360/5 = 72

360-72 = 288

288+360=648

648/360=9/5

And from the physical or six-fold symmetry:

360/5 =60

360-60=300

300-60=240

240/360=2/3

2/3 + 1 = 5/3

Five-thirds (yang) is the ratio of the salt metal potassium to the salt metal sodium. The 9/5 (yin) is the ratio of the molar mass of gold to the molar mass of silver and the ratio of the solar radius to the lunar orbital distance; The sun is gold in color and the moon is silver in color. Thus Antonio is an embodiment of yang (5/3) and we might guess his brother, Manolin, is an embodiment of yin (9/5) because it was when he told me he was in love with the moon that we discovered its connection to the yin, or 9/5. Manolin means little Manuel, and Manuel we might say is the embodiment of 15 because he introduced it to me as important to the universe (Discover And Contact by Ian Beardsley). The Earth rotates through 15 degrees in one hour. It may be that Manolin having matured now goes by the name of Manolo. Anyway, we have found the personification of 15, 9/5, and 5/3 upon which we built a Cosmology in my earlier work.

Ian Beardsley

April 22, 2014

CHAPTER 3

The Black Knight

The Black Night Satellite became know to the public on May 14 1954 in an article that appeared in the San Francisco Examiner. No one including NASA could explain its origins and it is thought by many to be an alien made satellite orbiting the Earth, monitoring us. It brings to mind the Monolith of 2001: A Space Odyssey. It turned out to be an alien sentinel placed on earth to monitor us and give us an evolutionary nudge when we needed it. Why did not NASA retrieve it with the space shuttle? My guess is they figured they did not know what they were dealing with. As we could learn from 2001: A Space Odyssey, the monolith should be approached with caution. The interesting thing is that A Ham Radio operator who received signals from it, decoded them and learned the satellite was 13,000 years old and came from the double star epsilon bootis. That is further interesting in that Our Cosmic Ancestors by Maurice Chatelain told of a story how there was some sort of echo in communication experiments. The echo was studied and its pattern if used on the televisions of the time came out to be a map of the constellation bootes as reported by the Scottish Astronomer Duncan Lunan, highlighting the same star that The Black Night has been reporting to as determined by the Ham Radio operator.

We learned of how the AE-35 Antenna of the Gypsy Shaman Manuel and its connection to the SETI Wow Signal in the constellation of Sagittarius was connected to him. When I was at the Gypsy caves, the son of Manuel, Antonio took a liking to the cowboy boots I wore. He showed me a pair, with great enthusiasm that he wore. Through his father, Manuel, we revealed the significance of 1.8 in the Universe. If we multiply 1.8 by three we have 5.4. Antonio's favorite meter to play in on the guitar was always three, both the fandango de huelva, and the fandango natural. 5.4 has the numbers of the 1954 of the first public article about the mysterious satellite. We might say as Manuel is associated with the SETI Wow Signal, or extraterrestrial signal from Sagittarius, his son Antonio is associated with the Ancient Alien Satellite called the black night because of the 5.4, the fandango and his interest in cowboy boots as epsilon bootes means the fourth brightest star in bootes and bootes sounds like plural for boots.

Ian Beardsley

April 23, 2014

CHAPTER 4

Syncopation

We can further think of syncopation at four over three or vice versa. Four times three is twelve and three quarters of twelve is (3/4)=36/4=9 which is the the ninth month or September which is the beginning of Autumn. Nine is the earth-sun separations of Saturn from the sun in its closest approach to the sun. Four thirds is (4/3)12=48/3=16. Sixteen months is four months from January is the month of April is the beginning of Spring. Thus (3/4)12 and (4/3)12 are Autumn and Spring are the equinoxes. The sixteen is important because it is the cycle of Indian Classical music that is capable of the most meaning. It is called Tin Taal.

Ian Beardsley

April 29, 2014

Bootes, Mars, Black Knight

Back in 2005 I in my book Decoding The Universe Two I wrote something that pertains to the Black Knight Satellite. Consider that the satellite is presumed to be extraterrestrial and is sending information to the constellation Bootes. I noticed that the closest star to us, Alpha Centauri, since it is the closest and I point out now that it is composed of a star just like the Sun, that it should be connected to the Earth. Well it is: the Earth is the third planet from the Sun and Alpha Centauri is the third brightest star in the sky. I then proceed to consider the largest planet, and most massive in the Solar System and it is Jupiter. If the above pattern repeats, it should connected to the brightest star in the sky. Well it is: Jupiter is the fifth planet from the Sun and Sirius is the fifth nearest star in the sky.

Let us now consider what Carl Sagan Said: to travel in space is to sail upon the cosmic ocean. Mars is the fourth planet from the Sun and is the best candidate for a body to colonize as it is solid (not gaseous) and is not too cold, or to warm. How appropriate! The brightest star in the constellation Bootes is Arcturus and it is the fourth brightest star in the sky. Also Bootes means "The Boatman". Is the Black Knight Satellite then, since theory has it is from Bootes, guiding us towards setting sail upon the cosmic ocean to colonize Mars?

Ian Beardsley

May 2, 2014

Discover And Contact Two

CHAPTER 6

The Short And Short

The short and short of it is we have two sets:

$(9/5, 5/3, 2)$
$(1, 4, 7)$

They are the two vectors:

$A = \langle 9/5, 5/3, 2 \rangle$
$B = \langle 1, 4, 7 \rangle$

They lay in a plane and the normal to the plane is their cross product:

$$A \times B = \begin{vmatrix} \vec{i} & \vec{j} & \vec{k} \\ 9/5 & 5/3 & 2 \\ 1 & 4 & 7 \end{vmatrix} = \left(\frac{5}{3}7 - 2(4) \right)\vec{i} + \left(2(1) - \frac{9}{5}7 \right)\vec{j} + \left(\frac{9}{5}4 - \frac{5}{3}(1) \right)\vec{k}$$

$$= \frac{11}{3}\vec{i} - \frac{53}{5}\vec{j} + \frac{83}{15}\vec{k}$$

The magnitude of the normal vector is:

$$\left| \left(\frac{11}{3} \right)^2 + \left(\frac{53}{5} \right)^2 + \left(\frac{83}{15} \right)^2 \right| = \frac{7039}{45} = 156.4$$

$$\sqrt{156.4} = 12.5$$

$$|A \times B| = AB\sin\theta$$

$$A = \sqrt{\left(\frac{9}{5} \right)^2 + \left(\frac{5}{3} \right)^2 + 2^2} = \sqrt{3.24 + 2.78 + 4} = \sqrt{10.02} = 3.165$$

$$B = \sqrt{1^2 + 4^2 + 7^2} = \sqrt{66} = 8.124$$

$$\sin\theta = \frac{12.5}{25.71246} = 0.4861$$

$$\theta = 29°$$

We find where the normal vector to the plane that contains the vectors A and B points on the celestial sphere by converting the cross product of A and B to right ascension and declination:

$a = 11/3$

$b = -53/5$

$c = \sqrt{\left(\dfrac{11}{3}\right)^2 + \left(\dfrac{53}{5}\right)^2} = \sqrt{13.44 + 112.36} = \sqrt{125.8} = 11.2$

$d = \dfrac{83}{15}$

$\tan\alpha = \dfrac{b}{a} = -\dfrac{55}{158} = -0.3459$

$\alpha = -19.0805° = RA$

$\tan\beta = \dfrac{83}{15} + 11.2 = 0.4940$

$\beta = 26.289° = dec$

$0.289(60) = 17.34$

$0.34(60) = 20.4$

$\beta = 26°17'20.4'' = dec$

$-\dfrac{19\deg}{15\deg/hour} = -1.267\,hours = RA$

$24\,hours - 1.267\,hours = 22.733\,hours = RA$

$(0.733)(60) = 43.98\min$

$(0.98\min)(60\sec) = 58.8\sec$

$RA = 22^h43^m59^s$

We now turn to our sky chart and find the brightest star closest to these coordinates is in the constellation Pegasus and is Mu Pegasus, which has the coordinates:

$22^h50^m42.07^s$

$24°40'32.5''$

Also, close to this star is Lambda Pegasus. 12 exoplanets have been found in Pegasus. The first to orbit a star like the Sun was found in the constellation orbiting 51 Pegasi, in 1995. The first exoplanet was found in 1992 orbiting a pulsar.

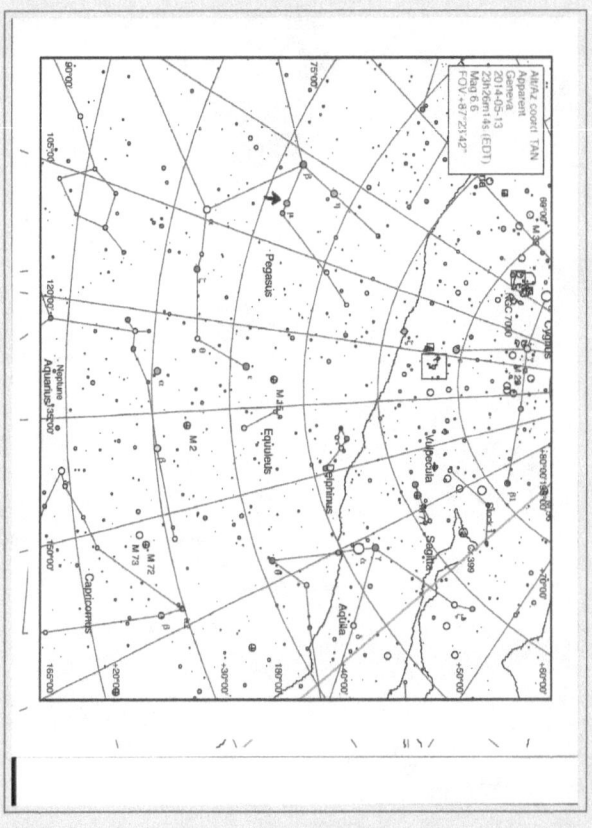

CHAPTER 7

Notes: May 07 2014

Zeta Reticuli

Lazar says they have been with us since we we were semian and that they have made some 60 odd modifications of our genome.

Bob Lazar MJ 12

No ET craft or Technology at area 51, They don't have the clearance. S4 was made for that.

Back engineered extraterrestrial craft. ET AUTOPSY He saw the et looked just like in lore. The big head, big eyes, thin body. typical grey, could not ascertain height from photo. The craft were just like reported ufos; disk shaped. Operate generating gravity waves. He was told ets came from zeta reticuli star system. Apparently some were on the base working with them.

When on considers
the coordinates of
Zeta Reticuli, they
find when converting
hours to degrees we
have about 45° for right ascension
and the declination is close
to 60°. These are important   60° in
angles in the special triangles   the negative
45 - 45 - 90, and 60-30-90.   direction

Zeta Reticuli is a double star
system with both components
like the sun. It is only about
400 ly from us.

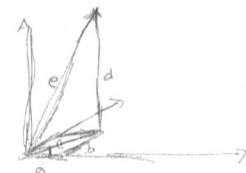

$\tan \frac{e}{a} = 45^\circ = RA$

$\tan \frac{d}{c} = 60^\circ = \text{declination}$

$c^2 = a^2 + b^2$

$e^2 = c^2 + d^2$

Coordinates of Zeta Reticuli

$3^h \ 18^m$ Approximately

$-62^\circ \ 18^m$ 45° and 60°

These are the two of three
regular tessellators.

$$\tan 45^\circ = \frac{b}{a}$$

$$\tan 60^\circ \quad \frac{d}{c}$$

$$c^2 = a^2 + b^2$$

$$a = b$$

$$\frac{b}{a} = 1$$

$$\frac{d}{c} = 1.732 = \frac{433}{250}$$

$$d = 433 \quad c = 250$$

$$(250)^2 = 2b^2$$

$$\frac{62,500}{2} = b^2 = 31,250$$

$$b = 176.7766953 \approx 177 = a$$

The vector pointing to zeta reticuli

is $\langle a, b, d \rangle = \langle 177, 177, -433 \rangle$

Ian Beardsley
May 09, 2014

We Can Make The Following
Table

$\dfrac{9}{5}$	$\dfrac{5}{3}$	15
Yin	Yang	2
mandin	antonio	Manuel
Soleares	Fandango	~~Hose~~ collector/ Shaman
~~list~~ organic	physical	one hour of earth rotation
Epsilon Boötes	Black kinght satellite	Wow! Signal Sagittarius

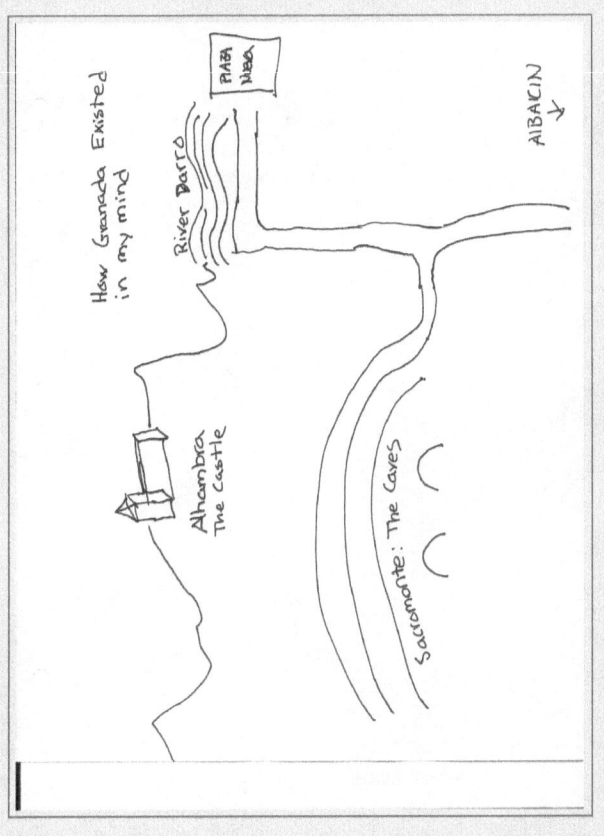

Moving On

We have said, whether or not my telling of the story is accurate, that we at least think there is a 13,000 year old satellite orbiting the Earth that comes from the Epsilon Bootis star system. We have said the physicist Bob Lazar worked for the government on a base that back engineered extraterrestrial spacecraft from the Zeta Reticuli star system. We have made a table that interconnects key Gypsies with key symbolic elements. We have made a map of Granada. It is now time to capture the spirit of Granada by presenting some verse from songs from Spain. Whether or not these words are my correct hearing of songs by Camaron de la Isla, does not matter, because this is what I hear and it matches the spirit of Granada to me:

La Gitana Morena

Nacio en Albaicin

Si Yo que a mi no,… par el tricin volver

Para trar

That is:

The Dark Gypsy Girl

She was born in Albaicin

If it doesn't matter to me

Three times he will return

For him returning

And:

Porque la Reina Sumara

Con su canaster Gitana

Y una bien parecia

Aye Sumara, Reina De La Moreria

That is:

Because the Queen Sumara

With her Gypsy Basket

And one well like it

Aye, Sumara,... Queen of the Moorness

Ian Beardsley

May 12, 2014

All Comes Together

In Discover And Contact we began with Manuel, the Chovihano, or Gypsy Shaman. We began with how he made me into the AE-35 Antenna, which was the unit in the movie 2001: a Space Odyssey that was reported to fail by the ship Discovery computer HAL to astronaut Dave Bowman, within 72 hours. We noted that making me into the AE-35 antenna made me telepathic with Manuel, which was not a real telepathy, but psychological trick induced when he showed me his collection of 15 rubber hoses in his cave in Granada, Spain. 72 is important because as it would turn out the precession of the Earth Equinoxes are one degree in 72 years. The number 15 is important because the Earth rotates through 15 degrees in one hour.

But it becomes more interesting when I show that Gold to Silver in molar mass is nine to five and so is the solar radius to the lunar orbital radius, and, the sun is Gold in color, the moon silver. This results in the Neptune Equation, which has the multiplier of 7.2 from Venus Orbit in astronomical units and Venus is a failed Earth, and Hal reports the AE-35 will fail in 72 hours.

As if this was not enough, we showed that 9 to 5, which is one point eight, describes the digits after the decimals both in pi added to the golden ratio and in pi added to Euler's number and, 9/5 results in one of three equations that make a plane in space whose normal vector points to the constellation Sagittarius, where SETI, the program for the Search for Extraterrestrial Intelligence detected a signal on August 15, 1977 which had all of the characteristics of being a signal from off earth intelligence. The signal lasted exactly 72 seconds: the same 72 of the 72 hours before the AE-35 would fail aboard Discovery, the same 72 that is the 0.72 Astronomical Units of Venus, a failed Earth, from the Sun, and the same 72 years for the precession of the Earth Equinox.

7.2 occurs in my Neptune equation. Neptune is Poseidon, the Greek God of the Sea. We can say both that 2001: A Space Odyssey is a space odyssey version of the Ancient Greek work Illiad and the Odyessy by Homer, and that it is a play on the idea that when we went into space, it was not unlike the early Odysseys upon the ocean in the Homer's Illiad and the Odyssey. My Neptune equation is connected to the planet Neptune, which is named for the Greek God of the ocean: When you set sail upon the sea, you have appeal to the God of the ocean Neptune (Poseidon, Greek), and now we must appeal to Neptune the planet to travel safely through space.

Now we have delved into other experiences in the Sacromonte (Gypsy Caves), to see if we can find more. We have. We created two sets of three numbers, found their cross product and that it points to the constellation Pegasus. As it would turn out, Pegasus is the constellation where the first exoplanet was discovered to orbit a star like the Sun. Before that we looked at epsilon bootis because the constellation bootes is the boatman, and as well their seems to be some indication that there is a thirteen thousand year old satellite orbiting the earth reporting to that constellation. Finally, since government physicist, Bob Lazar, has disclosed that he was back engineering alien spacecraft and was told the craft came from the star Zeta Reticuli star system, we looked at that star. We calculated that if you convert its Right ascension to degrees, it comes out to be very close to 45 and we also note its declination is almost -60 degrees; these are important angles in special regular polygons that tesselate and and allow us to calculate trignometric functions of angles. But most interestingly, we say now, is that the constellation to which Zeta Reticuli belongs is Reticulum, which was named by an astronomer Nicolas

Louis de Lacaille, to commemorate the reticle in his telescope eyepiece in the eighteenth century. He was using the eyepiece to define the constellation as a bust of Columbus. It was the Italian Columbus who was funded by Spain (Isabella) to discover a westerly route to India based on the idea that the world was round, but landed in America, discovering it.

Sometime after the Gypsy Shaman, Manuel of the Spain of Isabella, made me the AE-35 Antenna of a spaceship, not sea ship (like Columbus sailed), called Discovery, and set me on a course for contemplating space travel, I was intercepted by the Italy of Columbus, when I married a woman from Italy and went there with her several times, which set me on a course to discover that the nature of the universe and mathematical constants are connected to the central extraterrestrial activity that seems to be occurring here on Earth. Concerning my work Discover and Contact Two Cristopher Columbus died on my Birthday, May 20 and where I went in Spain was Granada, which lead to my discoveries, and the tomb of Isabella la Catolica is in Granada. She funded the voyage of Columbus.

Ian Beardsley

May 16, 2015

Review

All That Can Be Said

    They originated in the Far East, and passed some through the south, and others through the north over a period of A Thousand Years through deserts. They became Artists, and camped where there was no one else, and went unseen and unknown during all that time, only to unite and settle at Land's End, in neighborhoods, barrios, or quarters of western construct, and to let out only one verse of one of their poets, who is anonymous: "If there is someone in the street, he is familiar with it. If there is someone in the street, he knows him."

Chapter 1

AE-35

I wrote a short story last night, called Gypsy Shamanism and the Universe about the AE-35 unit, which is the unit in the movie and book 2001: A Space Odyssey that HAL reports will fail and discontinue communication to Earth. I decided to read the passage dealing with the event in 2001 and HAL, the ship computer, reports it will fail in within 72 hours. Strange, because Venus is the source of 7.2 in my Neptune equation and represents failure, where Mars represents success.

Ian Beardsley

August 5, 2012

It must have been 1989 or 1990 when I took a leave of absence from The University Of Oregon, studying Spanish, Physics, and working at the state observatory in Oregon -- Pine Mountain Observatory—to pursue flamenco in Spain.

The Moors, who carved caves into the hills for residence when they were building the Alhambra Castle on the hill facing them, abandoned them before the Gypsies, or Roma, had arrived there in Granada Spain. The Gypsies were resourceful enough to stucco and tile the abandoned caves, and take them up for homes.

Living in one such cave owned by a gypsy shaman, was really not a down and out situation, as these homes had plumbing and gas cooking units that ran off bottles of propane. It was really comparable to living in a Native American adobe home in New Mexico.

Of course living in such a place came with responsibilities, and that included watering its gardens. The Shaman told me: "Water the flowers, and, when you are done, roll up the hose and put it in the cave, or it will get stolen". I had studied Castilian Spanish in college and as such a hose is "una manguera", but the Shaman called it "una goma" and goma translates as rubber. Roll up the hose and put it away when you are done with it: good advice!

So, I water the flowers, rollup the hose and put it away. The Shaman comes to the cave the next day and tells me I didn't roll up the hose and put it away, so it got stolen, and that I had to buy him a new one.

He comes by the cave a few days later, wakes me up asks me to accompany him out of The Sacromonte, to some place between there and the old Arabic city, Albaicin, to buy him a new hose.

It wasn't a far walk at all, the equivalent of a few city blocks from the caves. We get to the store, which was a counter facing the street, not one that you could enter. He says to the man behind the counter, give me 5 meters of hose. The man behind the counter pulled off five meters of hose from the spindle, and cut the hose to that length. He stated a value in pesetas, maybe 800, or so, (about eight dollars at the time) and the Shaman told me to give that amount to the man behind the counter, who was Spanish. I paid the man, and we left.

I carried the hose, and the Shaman walked along side me until we arrived at his cave where I was staying. We entered the cave stopped at the walk way between living room and kitchen, and he said: "follow me". We went through a tunnel that had about three chambers in the cave, and entered one on our right as we were heading in, and we stopped and before me was a collection of what I estimated to be fifteen rubber hoses sitting on ground. The Shaman told me to set the one I had just bought him on the floor with the others. I did, and we left the chamber, and he left the cave, and I retreated to a couch in the cave living room.

Chapter 2

Gypsies have a way of knowing things about a person, whether or not one discloses it to them in words, and The Shaman was aware that I not only worked in Astronomy, but that my work in astronomy involved knowing and doing electronics.

So, maybe a week or two after I had bought him a hose, he came to his cave where I was staying, and asked me if I would be able to install an antenna for television at an apartment where his nephew lived.

So this time I was not carrying a hose through The Sacromonte, but an antenna.

There were several of us on the patio, on a hill adjacent to the apartment of The Shaman's Nephew, installing an antenna for television reception.

Chapter 3

I am now in Southern California, at the house of my mother, it is late at night, she is a asleep, and I am about 24 years old and I decide to look out the window, east, across The Atlantic, to Spain. Immediately I see the Shaman, in his living room, where I had eaten a bowl of the Gypsy soup called Puchero, and I hear the word Antenna. I now realize when I installed the antenna, I had become one, and was receiving messages from the Shaman.

The Shaman's Children were flamenco guitarists, and I learned from them, to play the guitar. I am now playing flamenco, with instructions from the shaman to put the gypsy space program into my music. I realize I am not just any antenna, but the AE35 that malfunctioned aboard The Discovery just before it arrived at the planet Jupiter in Arthur C. Clarke's and Stanley Kubrick's "2001: A Space Odyssey". The Shaman tells me, telepathically, that this time the mission won't fail.

Chapter 4

I am watching Star Wars and see a spaceship, which is two oblong capsules flying connected in tandem. The Gypsy Shaman says to me telepathically: "Dios es una idea: son dos". I understand that to mean "God is an idea: there are two elements". So I go through life basing my life on the number two.

Once one has tasted Spain, that person longs to return. I land in Madrid, Northern Spain, The Capitol. The Span-iards know my destination is Granada, Southern Spain, The Gypsy Neighborhood called The Sacromonte, the caves, and immediately recognize I am under the spell of a Gypsy Shaman, and what is more that I am The AE35 Antenna for The Gypsy Space Program. Flamenco being flamenco, the Spaniards do not undo the spell, but reprogram the instructions for me, the AE35 Antenna, so that when I arrive back in the United States, my flamenco will now state their idea of a space program. It was of course, flamenco being flamenco, an attempt to out-do the Gypsy space pro-gram.

Chapter 6

I am back in the United States and I am at the house of my mother, it is night time again, she is asleep, and I look out the window east, across the Atlantic, to Spain, and this time I do not see the living room of the gypsy shaman, but the streets of Madrid at night, and all the people, and the word Jupiter comes to mind and I am about to say of course, Jupiter, and The Spanish interrupt and say "Yes, you are right it is the largest planet in the solar system, you are right to consider it, all else will flow from it."

I know ratios, in mathematics are the most interesting subject, like pi, the ratio of the circumference of a circle to its diameter, and the golden ratio, so I consider the ratio of the orbit of Saturn (the second largest planet in the solar system) to the orbit of Jupiter at their closest approaches to The Sun, and find it is nine-fifths (nine compared to five) which divided out is one point eight (1.8).

I then proceed to the next logical step: not ratios, but proportions. A ratio is this compared to that, but a proportion is this is to that as this is to that. So the question is: Saturn is to Jupiter as what is to what? Of course the answer is as Gold is to Silver. Gold is divine; silver is next down on the list. Of course one does not compare a dozen oranges to a half dozen apples, but a dozen of one to a dozen of the other, if one wants to extract any kind of meaning. But atoms of gold and silver are not measured in dozens, but in moles. So I compared a mole of gold to a mole of silver, and I said no way, it is nine-fifths, and Saturn is indeed to Jupiter as Gold is to Silver.

I said to myself: How far does this go? The Shaman's son once told me he was in love with the moon. So I compared the radius of the sun, the distance from its center to its surface to the lunar orbital radius, the distance from the center of the earth to the center of the moon. It was Nine compared to Five again!

I had found 9/5 was at the crux of the Universe, but for every yin there had to be a yang. Nine fifths was one and eight-tenths of the way around a circle. The one took you back to the beginning which left you with 8 tenths. Now go to eight tenths in the other direction, it is 72 degrees of the 360 degrees in a circle. That is the separation between petals on a five-petaled flower, a most popular arrangement. Indeed life is known to have five-fold symmetry, the physical, like snowflakes, six-fold. Do the algorithm of five-fold symmetry in reverse for six-fold symmetry, and you get the yang to the yin of nine-fifths is five-thirds.

Nine-fifths was in the elements gold to silver, Saturn to Jupiter, Sun to moon. Where was five-thirds? Salt of course. "The Salt Of The Earth" is that which is good, just read Shakespeare's "King Lear". Sodium is the metal component to table salt, Potassium is, aside from being an important fertilizer, the substitute for Sodium, as a metal component to make salt substitute. The molar mass of potassium to sodium is five to three, the yang to the yin of nine-fifths, which is gold to silver. But multiply yin with yang, that is nine-fifths with five-thirds, and you get 3, and the earth is the third planet from the sun.

I thought the crux of the universe must be the difference between nine-fifths and five-thirds. I subtracted the two and got two-fifteenths! Two compared to fifteen! I had bought the Shaman his fifteenth rubber hose, and after he made me into the AE35 Antenna one of his first transmissions to me was: "God Is An Idea: There Are Two Elements".

It is so obvious, the most abundant gas in the Earth Atmosphere is Nitrogen, chemical symbol 15!